THE ELEMENTS

Potassium

Chris Woodford

BENCHMARK BOOKS

MARSHALL CAVENDISH
NEW YORK

Benchmark Books
Marshall Cavendish
99 White Plains Road
Tarrytown, New York 10591

www.marshallcavendish.com

Library of Congress Cataloging-in-Publication Data

Woodford, Chris.
Potassium / Chris Woodford.
p.   cm. – (The elements)
Includes index.
Summary: Describes the characteristics, sources, and uses of the element potassium.
Contents: What is potassium? – Where is it found? – How was it discovered? – Where
does it come from? – Special characteristics – How potassium reacts – Soaps and
detergents – Potassium and plants – Potassium in you – Potassium and explosives –
Periodic table – Chemical reactions
ISBN 0-7614-1463-0
1. Potassium—Juvenile literature. [1. Potassium.] I. Title. II. Series.
QD181.K1 W66 2002
546'.383—dc21
2002018556

Printed in Hong Kong

Picture credits
Front cover: Jeff Hunter at Image Bank
Back cover: Jerry Mason at Science Photo Library
AKG: Deutsches Bergbau-Museum 8; Eric Lessing 25; Wilhelm Trautschold 20
Corbis: Brian Bailey 23; Lester V. Bergman 22; Ecoscene 21; Eric Hausman 14;
Hanan Isachar 6; Buddy Mays 9
Du Pont: 26
G.S.F. Picture Library: 10
Hulton Archive: 15, 16
Image Bank: Larry Keenan iii, 13; Jeff Hunter 27
Science & Society Picture Library: Science Museum 7
Science Photo Library: Astrid & Hanns-Frieder 4 (top); Jerry Mason 12;
David Taylor 30; Charles D. Winters 11
Still Pictures: Joerg Boethling 5; Ron Giling 19
USDA: Scott Bauer/ARS i, 24 (top), 24 (bottom); Joe Larson/NRCS 4 (bottom);
Tim McCabe/NRCS 18

Series created by Brown Partworks Ltd.
Designed by Sarah Williams

Contents

What is potassium?

Potassium is one of the most common elements in Earth's crust, but you are unlikely to find it as an element in nature. This soft, silvery-white metal is so reactive that it almost always exists combined with other elements as compounds. All living things—from plants to people—need potassium to stay healthy. Potassium compounds have many other important uses. They help matches to burn and give fireworks their bright, colorful explosions. They are also found in breathing apparatus, cotton dyes, liquid soaps, and photographic chemicals.

Potassium-rich fertilizers ensure that the vegetables we eat, such as carrots, are nutritious and tasty. All living things need potassium to stay healthy.

Soft, silvery chunks of potassium are so reactive that they have to be covered with a film of oil to prevent the metal from reacting with oxygen in the air.

The potassium atom

Everything in the universe consists of tiny particles called atoms. Atoms are made up of even smaller particles called electrons, neutrons, and protons. The neutrons and protons cluster together in the nucleus at the center of each atom. The electrons

DID YOU KNOW?

WHY POTASSIUM HAS THE SYMBOL K
Most chemical elements have symbols that come directly from the first few letters of their name. For example, hydrogen has the symbol "H," helium has the symbol "He," and carbon has the symbol "C." Potassium does not follow this rule. It has the symbol "K," which comes from the word *kalium*. This is the Latin name for a substance called potash—a potassium compound that was used by people long before the element potassium was discovered.

orbit the nucleus in layers called electron shells. There are 19 negatively charged electrons orbiting the nucleus of each potassium atom. The number of electrons and protons in an atom is always the same, so potassium has 19 positively charged protons in the nucleus. Neutrons are about the same size as protons but have no electrical charge. Most potassium atoms have 20 neutrons in the nucleus.

Forming chemical bonds

Atoms are most stable if the outer electron shell is full. Some atoms share electrons with other elements to make them stable. The easiest way for a potassium atom to become stable is by losing a single electron from the outer electron shell. During a

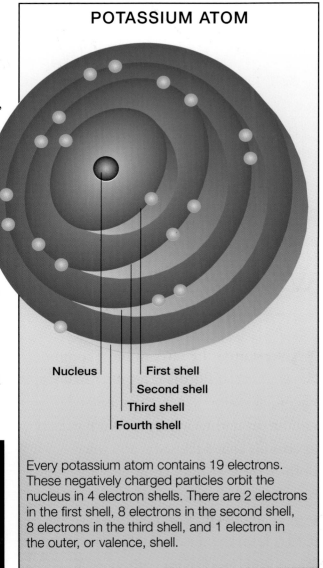

POTASSIUM ATOM

Nucleus | First shell
Second shell
Third shell
Fourth shell

Every potassium atom contains 19 electrons. These negatively charged particles orbit the nucleus in 4 electron shells. There are 2 electrons in the first shell, 8 electrons in the second shell, 8 electrons in the third shell, and 1 electron in the outer, or valence, shell.

chemical reaction, the electron transfers over to another element so that the potassium atom forms a chemical bond with the other element. Potassium loses this electron very easily, so it is highly reactive.

A factory worker in Tamil Nadu, India, uses a potassium compound to fix a blue dye to cotton fabric.

Where is it found?

A potash evaporation mine on the coast of the Dead Sea in Sodom, Israel.

Potassium is the seventh most common element in Earth's crust after oxygen, silicon, aluminum, iron, calcium, and sodium. In nature, most potassium occurs as crystals called feldspars and micas. Both of these crystals contain potassium combined with aluminum, silicon, and oxygen. Rainwater and carbon dioxide (CO_2) from the air gradually turn feldspars into potash (potassium carbonate; K_2CO_3). Potash supplies most of the potassium used by plants. Nature is very good at recycling chemical elements. Some of the largest natural deposits of potassium compounds on Earth today come from the remains of prehistoric plant life.

Other naturally occurring potassium compounds include silvite (potassium chloride; KCl), silvinite (a mineral formed from sodium chloride [NaCl] and potassium chloride), langbeinite (a sulfate of potassium and magnesium), and carnallite (a mineral formed from potassium chloride, magnesium chloride [$MgCl_2$], and water [H_2O]).

DID YOU KNOW?

MINING POTASSIUM

Potassium deposits are found all over the world. Mines in the Russian Federation produce about one third of the world's annual supply, with Canada and Germany producing around one quarter each. Other countries with smaller deposits include France, the United States, Israel, and Jordan. Some of the biggest deposits in the United States are found at Searles Lake, California, and Carlsbad, New Mexico.

How was it discovered?

For many years, most chemists thought that potash was a chemical element. When the so-called "founder of modern chemistry," Frenchman Antoine-Laurent Lavoisier (1743–1794), published a list of all the known chemical elements in 1789, he left out potash because he thought that it was a compound. In 1807, British chemist Sir Humphry Davy (1778–1829) proved Lavoisier to be right. Davy revolutionized the study of chemistry when he found a new way of isolating chemical elements from their compounds. Davy passed an electrical current through chemical compounds to make them break apart. Davy first used this method—later called electrolysis—to separate potassium from potash. Davy also used electrolysis to isolate the elements sodium, calcium, strontium, barium, and magnesium.

The samples of the metals potassium, calcium, lithium, iron, and strontium shown in this picture were purified using electrolysis—a technique pioneered by Sir Humphry Davy.

Where does potassium come from?

There is little demand for potassium as a metal because it is so highly reactive that it instantly reacts with water and oxygen in the atmosphere to form potassium compounds. Sodium behaves in exactly the same way as potassium and is also less expensive. Because of this, sodium often replaces potassium for most uses. The small amount of potassium that is used in industry is extracted from the minerals silvite and carnallite, both of which contain the compound potassium chloride. Unlike many other metals, there is little chance of these potassium sources

An artist's impression of work at the large Stassfurt mines, Germany, in 1900. These mines form the largest deposits of potassium minerals in the world.

running out. The vast Stassfurt mines in Germany, which were discovered in 1856, are estimated to contain supplies that will last for several thousand years.

Extracting potassium

In industry, potassium is made by reacting potassium chloride with sodium. During the reaction, the chemical bonds between the potassium and chlorine atoms break. The chlorine atoms then combine with the sodium atoms to form sodium chloride, or common salt ($NaCl$). The potassium metal is left alone.

This reaction only works if the reactants are heated to very high temperatures. In one part of the chemical plant where potassium is extracted, sodium chloride is

---

DID YOU KNOW?

THE GREAT POTASSIUM SHORTAGE
During World War I (1914–1918), when many countries were at war with Germany, supplies of potassium minerals from the Stassfurt mines in Germany were no longer available. Chemists had to find alternative sources of potassium. One source came from a potassium-rich seaweed called kelp. In 1916, a large factory was built on the coast of California to grow kelp, burn it, and recover the potassium. Another source of potassium came from waste products of the sugar-refining industry. Eight plants were built in the United States during World War I to recover potassium from this alternative source.

heated so that the sodium escapes as a vapor. Pumps force the sodium vapor up into a reaction vessel called an exchange column. In another part of the plant, potassium chloride is heated to about 1550° F (843° C). The potassium chloride molecules melt and then pumps push the melted molecules to the top of the column. As the melted potassium chloride travels down the column it meets the sodium vapor that is traveling up. The chlorine atoms bond with the sodium, forming potassium vapor and sodium chloride. The sodium chloride is removed at the bottom of the exchange column, while the potassium vapor is collected at the top of the vessel. The resulting potassium is nearly 100 percent pure.

A huge chunk of the potassium mineral silvite is found at a mine in Carlsbad, New Mexico.

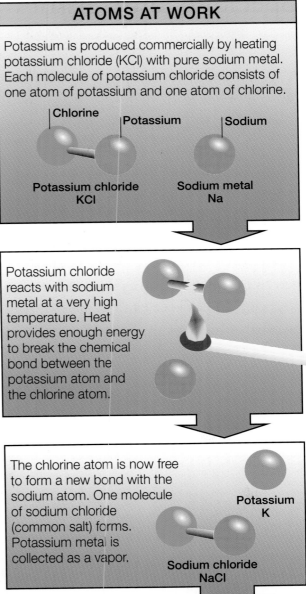

ATOMS AT WORK

Potassium is produced commercially by heating potassium chloride (KCl) with pure sodium metal. Each molecule of potassium chloride consists of one atom of potassium and one atom of chlorine.

Chlorine   Potassium   Sodium

Potassium chloride
KCl

Sodium metal
Na

Potassium chloride reacts with sodium metal at a very high temperature. Heat provides enough energy to break the chemical bond between the potassium atom and the chlorine atom.

The chlorine atom is now free to form a new bond with the sodium atom. One molecule of sodium chloride (common salt) forms. Potassium metal is collected as a vapor.

Potassium
K

Sodium chloride
NaCl

The chemical reaction that takes place when sodium metal reacts with potassium chloride can be written like this:

$KCl + Na \rightarrow NaCl + K$

This equation tells us that one molecule of potassium chloride reacts with one atom of sodium to form one molecule of sodium chloride and one atom of potassium.

Special characteristics

POTASSIUM FACTS

○ Potassium melts at 147° F (64° C) and boils at 1398° F (759° C). At normal room temperature and atmospheric pressure, potassium is a solid.

○ On the Mohs' hardness scale, potassium has a value of 0.4, which means that it is even softer than talc.

○ Like most metals, potassium conducts both heat and electricity extremely well.

Potassium is one of the alkali metals—a number of elements that make up Group I on the left-hand side of the periodic table. Like other alkali metals, potassium has some unusual properties. For example, the metal is so soft that it can be cut with a knife. Its melting point is lower than the boiling point of water.

The chemistry of the alkali metals is determined by the arrangement of their electrons. The electrons of the alkali metals are farther away from the nucleus than in most other atoms, so the atoms of pure potassium metal are held together very weakly. This means that it does not take much to break potassium atoms apart. That is why potassium is soft (you can break apart the atoms with a knife) and why it has a low melting point (heat easily breaks the atoms apart).

Potassium is soft enough to be cut with a knife. The brown film on the surface of the metal is oil. Potassium is stored under oil to prevent the violent reaction of potassium with water and oxygen.

How potassium reacts

If you drop a small piece of potassium metal into a bowl of water, a violent reaction will take place. The potassium fizzes across the surface of the water, producing potassium hydroxide (KOH) and hydrogen gas (H_2). So much heat is produced that the hydrogen gas ignites, burning with a bright lilac flame. Potassium also reacts violently with some acids. A reaction between potassium and a halogen like bromine can be explosive. Potassium also reacts with oxygen in the air, forming potassium monoxide (K_2O) and potassium peroxide (K_2O_2).

Carbon compounds

Potassium does not react with carbon to form potassium carbide. Instead, potassium forms what is known as a "solid solution" with potassium atoms trapped inside the carbon crystal structure. However, potassium does react with a number of organic (carbon-containing) compounds such as alcohols and hydrocarbons. Hydrocarbons are compounds consisting of both carbon and hydrogen.

Common potassium compounds

Reactions involving potassium produce a number of important compounds. Potassium chloride, widely found in

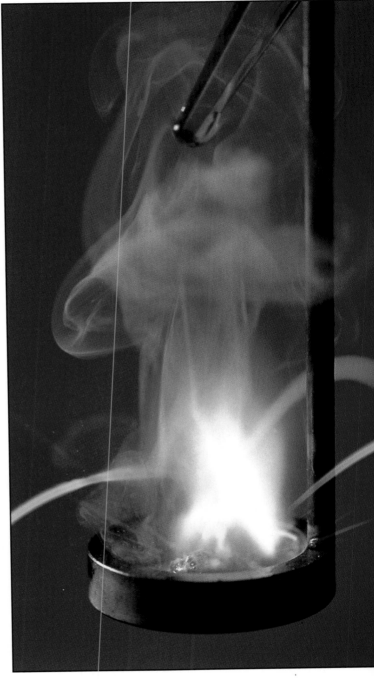

Water drips from the tip of a pipette into a dish containing potassium metal. The reaction yields potassium hydroxide (KOH) and hydrogen gas (H_2). The hydrogen ignites instantly, burning with a lilac-colored flame.

11

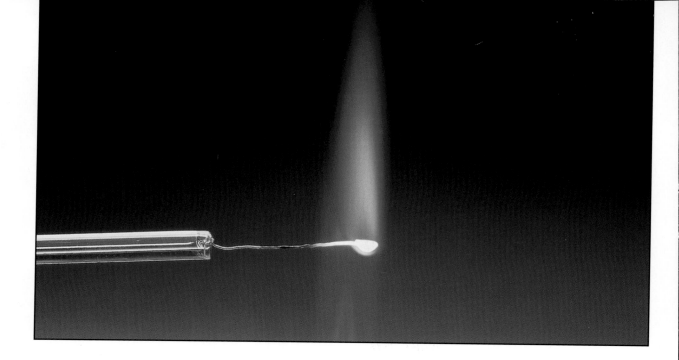

A flame test is a simple way to test for the presence of certain metals in a compound. The lilac color of this flame indicates that potassium is present in the compound being tested.

nature, is a good fertilizer. Potassium nitrate (KNO_3)—also called niter or saltpeter—is another good fertilizer. It is produced by reacting potassium chloride and sodium nitrate ($NaNO_3$). Potassium nitrate and potassium chlorate ($KClO_3$) also help matches to burn brightly. When the potassium compounds are heated they release oxygen, which fuels the flame.

Some potassium alloys—mixtures of two or more metals—are also important substances. For example, an alloy of sodium and potassium is used as a catalyst (a substance that speeds up a chemical reaction) in the manufacture of lard, which is sometimes used in cooking.

Why potassium is so reactive

All the alkali metals are very reactive, but each one is slightly more reactive than the one directly above it in the periodic table. So sodium is more reactive than lithium and potassium is more reactive than sodium. But why does this happen?

The alkali metals have a large atomic radius—their electrons spread out farther from the nucleus than the electrons of other elements. Electrons are attracted to protons in the nucleus. The larger the atomic radius the weaker the attractive force will be. This is why electrons farther from the nucleus are lost more easily during a reaction. As you go down Group I in the periodic table, the elements have larger atomic radii. Potassium has a larger atomic radius than sodium and loses electrons more easily. This is why potassium is more reactive than sodium.

Redox reactions

When potassium reacts with oxygen in air, potassium atoms lose electrons and oxygen atoms gain electrons. Losing electrons during a chemical reaction is called oxidation. Gaining an electron in a reaction is known as reduction. Oxidation and reduction always happen together. A redox reaction is a reaction where oxidation and reduction take place.

In the reaction between potassium and oxygen, potassium is oxidized and oxygen is reduced. Potassium is good at losing electrons and is called a reducing agent. Because oxygen accepts electrons readily it is called an oxidizing agent.

The reaction between red phosphorus in the strip on a matchbox and other substances in the match head, such as potassium nitrate (KNO_3) or potassium chlorate ($KClO_3$), causes matches to burn brightly.

ATOMS AT WORK

When potassium reacts with water, the result can be explosive. One molecule of water consists of two hydrogen atoms joined to one oxygen atom.

| Hydrogen | Oxygen | Potassium |

Water 2x H_2O   **Potassium 2x K**

When water comes into contact with potassium, the attraction between the potassium and the oxygen atoms in the water is so strong that one of the hydrogen atoms on each water molecule breaks free. The two hydrogen atoms then bond to form a molecule of hydrogen gas.

The reaction yields two molecules of potassium hydroxide and one molecule of hydrogen gas.

Hydrogen gas   **Potassium hydroxide**
H_2   **2x KOH**

The chemical reaction that takes place can be written like this:

$$2K + 2H_2O \rightarrow 2KOH + H_2$$

The number of atoms of each element is the same on each side of the equation, although the atoms have formed new combinations.

Soaps and detergents

A surgeon washes his hands with soap before an operation. Modern soaps contain antibacterial agents that help to kill harmful microorganisms.

Soaps and detergents are substances that help us to remove dirt from our clothes and our bodies. Potash has played an important role in the manufacture of soaps and detergents for thousands of years. Potassium carbonate and potassium hydroxide are still used to make certain liquid soaps and detergents.

Water and surface tension

Water is not always successful at cleaning up dirt. First, water molecules are simply not attracted to the dirt molecules. Water also has a property called surface tension. When water molecules are together, they are surrounded and attracted by other water molecules. At the surface of the water a tension is created as the water molecules are pulled together. This makes the water form droplets on your skin and clothes, which slows the wetting and cleaning processes. Substances called surfactants reduce the surface tension of water so that it can spread out and wet things more efficiently.

How soaps and detergents work

Soaps and detergents change the properties of water. Soap and detergent molecules have two distinct ends. One end of the molecule is similar to a water molecule and will be attracted to water molecules. The other end of the molecule is similar to molecules of dirt and will be attracted to dirt particles. Soaps and detergents act like a kind of adhesive between water and dirt.

When cleaning agents are added to water, they also reduce its surface tension. This makes the water spread out over our clothes or skin. It becomes easier for the soap or detergent molecules to grab hold of dirt and pull it away. Once the dirt particle has been pulled away, other soap or detergent molecules engulf the dirt and stop it from reattaching itself to the clothes or skin. They hold the dirt in the water until it can be washed away.

Making soaps and detergents

Soaps and detergents are made in similar ways. Detergents are more effective cleaning agents than soaps because they contain more than one surfactant.

An engraving shows workers stirring sunken vats of semisolid soap inside a soapmaking factory in 1771.

Generally, soaps and detergents are made by reacting a hot alkaline solution of potassium or sodium with a fat or oil. This reaction, called saponification, produces a smooth, semisolid salt of potassium or sodium (the soap) and a colorless liquid called glycerol (a waste product from the reaction). The fats and oils used in soapmaking come from a number of different sources. Corn oil, coconut oil, fish oils, and animal fats such as tallow—a hard fat made from the remains of sheep and cattle—are all used to make different soaps. Petrochemicals (chemicals that come from petroleum or natural gas) are also used to make some detergents.

In industry, soaps are made in large chemical plants called hydrolyzers. Melted fat and water are pumped continuously into the hydrolyzer. High-pressure steam hydrolyzes (splits) the fats and oils into fatty acids and glycerol. Fatty acids are long chains of carbon and hydrogen atoms with a carboxylic acid group (-COOH) at one end. The fatty acids are then purified by distillation and neutralized with the potassium or sodium alkalis to produce the

A vat of liquid soap is stirred at the Lever Brothers'
soap factory at Port Sunlight in Cheshire, Britain.
Soaps are made by treating fats and oils, or their
fatty acids, with a strong sodium or potassium alkali.

liquid soap or detergent. Perfumes and colors are then added. Most soap is then cooled, dried, and shaped into solid bars.

Sodium alkalis are used more frequently in the soapmaking industry because they are cheaper than potassium alkalis. Some liquid soaps and detergents, however, are still made using potassium alkalis such as potassium hydroxide.

The history of soap

No one can be sure who made the first soap. Simple methods of soapmaking may have been developed around 3500 B.C.E. by the Sumerians—an early civilization from Mesopotamia (present-day Iraq). Records indicate that later civilizations, including the Phoenicians and Romans, knew how to make soap from animal fats and potash. Soapmaking became much more common during the Middle Ages (c.500-1500 C.E.).

In 1789, French chemist Nicolas Leblanc (1742–1806) made a discovery that transformed soap from a luxury item into something everyone could afford. Before this time, soap was always made using potash as the alkali. In ancient times, the potash came from the ashes made by roasting plants in earthenware pots. Leblanc found a way of making another alkali, called soda (sodium carbonate; Na_2CO_3), using sodium chloride. Leblanc's new method made soapmaking much quicker and easier than before.

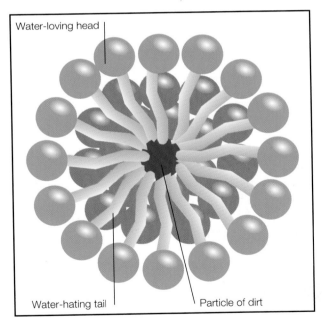

Water-loving head

Water-hating tail

Particle of dirt

Soap molecules have long "water-hating" tails and "water-loving" heads. In water, soap molecules surround a particle of dirt, with the water-hating tails in the center. The water then carries the dirt particle away with the soap.

Potassium and plants

Plants need sixteen major nutrients to grow well and stay healthy. Nine elements (nitrogen, phosphorus, potassium, carbon, hydrogen, oxygen, calcium, sulfur, and magnesium) are especially important. These elements are called macronutrients. The other seven (iron, boron, manganese, copper, chlorine, molybdenum, and zinc) are needed by plants in much smaller amounts and are called the minor elements. Most of the macronutrients come from the soil. As plants grow, however, they remove the elements from the soil. The elements must somehow be returned to the soil if healthy plants are to continue growing there in the future.

Potassium plays a very important role in keeping a plant healthy. The element is taken up from the soil by the roots in the form of potassium ions (K^+). These ions tend to encourage the "new growth" of a plant. They help to activate around 80 enzymes—catalysts that help to speed up chemical reactions in living things. Some enzymes are essential for plant growth. Others help the plant convert sugar into starch, which is the main ingredient in vegetables such as potatoes. Another vital role for potassium ions is to help the plant use water more efficiently. The ions regulate water loss from the plant's leaves

Liquid manure is pumped onto crops in northeastern Iowa. The use of natural fertilizers, such as manure, is probably as old as agriculture itself.

in a process called transpiration. If plants do not get enough potassium they die. This can be avoided by adding a potassium-rich fertilizer to the soil.

Many developing countries cannot afford to buy expensive agricultural machinery. These laborers from Bolivia are scattering artificial fertilizers by hand.

Fertilizers

Fertilizers are substances added to soil to make plants grow better. Long before the element was discovered, potassium-rich substances were used as fertilizers.

All fertilizers come in two main kinds—natural and artificial. Natural fertilizers include animal manure, seaweed, and the leaves that drop from trees in the fall. Scientists now understand the role different chemical elements play inside plants. The earliest farmers gained their understanding of fertilizers by trial and error. They accidentally discovered that they could make crops grow better by adding guano (bird droppings), wood ash (a source of potash), and other natural fertilizers.

An artist's impression of Baron Justus von Liebig's chemistry laboratory in Giessen, Germany.

The production of artificial fertilizers is now a multibillion dollar industry. Most artificial fertilizers contain nitrogen (N), phosphorus (P), and potassium (K). These are called NPK fertilizers. Different fertilizers are made to a variety of different "recipes." Farmers select the fertilizer to maximize the growth of a particular crop in a particular type of soil. Potassium chloride is the most important potassium compound used to make modern artificial fertilizers, but potassium nitrate, sulfate, and carbonate are also widely used.

DISCOVERERS

BARON JUSTUS VON LIEBIG

German chemist Baron Justus von Liebig (1803–1873) is widely considered to be the father of modern agricultural chemistry. Liebig was the first person to realize that plants took in inorganic substances, such as minerals from soil and gases in the air, to fuel their own growth. Liebig's studies of the soil and the elements that it contains added much to the modern understanding of how fertilizers work. However, Liebig is probably best known for inventing the Liebig condenser. This piece of laboratory equipment continues to be used in distillation—a process for purifying substances by turning them first into gases and then back into liquids.

DID YOU KNOW?

THE SENSITIVE PLANT

The sensitive plant *(Mimosa pudica)* takes its common name from the way its leaves tightly curl up at the slightest touch. When you touch a sensitive leaf, potassium ions move from one side of the leaf to the other. This causes water to flow from the leaves, too, making them fold up and droop in the same way as a balloon collapses when you release the air. Botanists (scientists who study plants) think that the sensitive plant developed this unusual characteristic to stop animals and insects from eating its leaves.

Environmental issues

Natural and artificial fertilizers are used throughout the world in vast quantities because they help farmers to produce more crops. But the overuse of fertilizers sometimes carries a considerable environmental price.

Eutrophication is an increase in the mineral content of natural water supplies. It is a natural process, but fertilizers can accelerate the effect. Eutrophication causes an excessive growth of algae (small water plants) in the water. These so-called "algal blooms" kill off marine and aquatic life. And the more often farmers disturb the soil by plowing fertilizers into it, the more it suffers from a type of damage called soil erosion. This can turn healthy, fertile soil into little more than dust and prevent it from growing crops in the future.

A canal running through a field of vegetable crops in Germany has been heavily polluted by the discharge of fertilizers into its waters.

Potassium in the body

Everyone needs to eat vitamins and minerals to stay healthy. Potassium is no exception. Potassium is used by the body in the form of ions (K^+). Potassium ions make up just 0.35 percent of the weight of an average human being. These amounts may seem small, but any more potassium ions in the body would be dangerous.

What potassium does

One of the most important uses of potassium ions is to help the body convert blood sugar into glycogen—a type of

POTASSIUM FACTS
● Potassium is known as a trace mineral, or trace element, because the body needs only very small amounts to function properly.
● Dieticians recommend that people should eat about 3,500 milligrams of potassium per day. This is about the same as eating five or six bananas.
● Potassium is found in most food because plants and animals need it to survive.
● Canned and frozen foods contain less potassium than fresh foods.

Glycogen cells in the liver. Like the energy stored in a battery, glycogen gives our muscles the energy to move our bodies without immediately becoming tired.

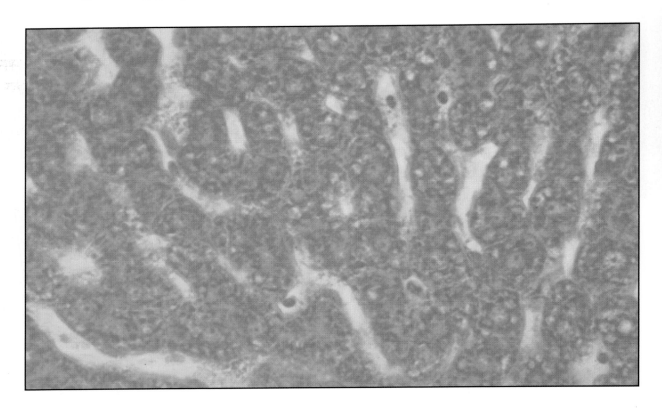

People who exercise heavily will sweat a lot and will need more potassium. They may have to supplement their diet with potassium-rich foods.

ions also support the balance of water between the body's cells and the fluids that surround them. As in plants, potassium ions activate enzymes. They also help to transmit electrical impulses through the nervous system.

Potassium is sometimes prescribed by physicians for people with hypertension (high blood pressure). Potassium ions relax large blood vessels, called arteries, and can lower blood pressure more effectively than some drugs. Many medical studies have shown that a diet rich in potassium may help to reduce the risk of developing heart disease and cancer.

How we get potassium

Since the human body needs only a small amount of potassium, a balanced diet is the best way to ensure you get enough potassium to stay healthy. Good sources of potassium include avocados, bananas, potatoes, tomatoes, and white meat. Red meat, apples, and broccoli provide smaller amounts of potassium. If you eat lots of fresh fruit and vegetables, you should not have to take mineral tablets to supplement the potassium in your diet.

Potassium deficiency

People whose bodies contain too little potassium may suffer from a condition called hypokalemia. Without potassium to help convert glycogen into blood sugar,

sugar stored in the liver. Potassium ions also help to maintain normal blood pressure, which is essential for keeping blood moving around the body. Potassium

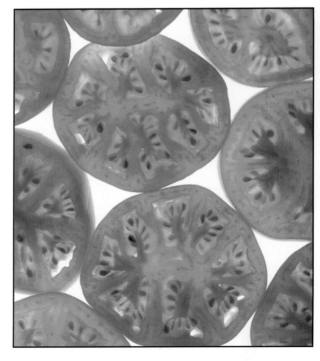

Potatoes (above) and tomatoes (right) are both excellent sources of potassium.

the first sign of hypokalemia is a general feeling of weakness. High blood pressure is another common sign of hypokalemia.

Besides a poor diet, various medical problems may also lead to a potassium deficiency. Loss of body fluids, kidney problems, and severe burns can all lead to a potassium deficiency. People who exercise a lot are also more likely to experience potassium deficiency because potassium ions are lost (along with other body salts) in sweat.

Potassium and explosives

Explosives are perhaps best known for their military applications, but they have many peacetime uses too. Explosives put the bang into fireworks, make matches ignite, and bring down the walls of quarries to yield supplies of important minerals. Potassium nitrate—originally the main chemical in gunpowder—is still an important ingredient in many modern explosives.

How explosives work

An explosion is a chemical reaction that generates a huge amount of fast-moving gas in a very short space of time. As the gas expands, it creates a "shock wave" that flattens everything in its path.

Different explosives use different chemical reactions. The earliest known explosive—gunpowder—was made mostly from potassium nitrate (saltpeter), with some charcoal and sulfur added to it. When the gunpowder ignites, it burns quickly to form potassium carbonate, potassium sulfate (K_2SO_4), and a huge amount of gas that creates the force of the explosion. Gunpowder can also be made without sulfur. Although this explosive is not as powerful, it does less damage to the inside of a gun.

This woodcut shows two German soldiers firing a cannon. Gunpowder was the only explosive available to fire such weapons until the invention of nitrogen-based explosives in the nineteenth century.

There are many different explosives, but they all fall into three main categories: primary explosives, low explosives, and high explosives. Primary explosives are not very powerful but will detonate if they are ignited. Low explosives, such as gunpowder, explode with relatively little force. Low explosives burn at a steady

In 1802, French-born industrialist E. I. du Pont de Nemours established a chemical plant to produce gunpowder near Wilmington, Delaware. This marked the successful beginning of the company Du Pont.

The history of explosives

Gunpowder was originally discovered by the Chinese around the ninth century C.E., but it was not used by Europeans until the thirteenth century. Often called black powder, gunpowder rapidly transformed medieval warfare. Until that time, battles had been fought with a mixture of hand-to-hand combat and simple mechanical artillery weapons such as catapults. Historians think that gunpowder was first used to fire weapons in the fourteenth century. From then on it was made in large quantities in countries such as England and Germany. Gunpowder remained the only explosive in common use until about the seventeenth century.

speed and are suitable for use in matches, some types of ammunition, and fireworks. The most powerful explosives are called high explosives. Most high explosives are nitrogen-based chemicals that release large amounts of energy when detonated. Nitroglycerin and trinitrotoluene (TNT) are both examples of high explosives.

DID YOU KNOW?

BLASTING

One of the most important peacetime uses for explosives is for blasting. Blasting is mainly used in quarries. Explosives reduce the rock face of a quarry into small fragments so that valuable mineral deposits can be extracted. Blasting has many other uses, however, from the construction of tunnels to the detonation of avalanches. The blasting process follows a series of steps. First, a pattern of holes is drilled into the rock, and an explosive charge and detonator is placed in each hole. The explosive is compacted into the hole, which is then filled with clay or rock fragments. Once the rock face is prepared, the explosive charge is detonated. When the area is safe, the broken rock fragments are cleared away.

Explosives developed rapidly in the mid-nineteenth century. In 1867, Swedish chemist Alfred Nobel (1833–1896) made a nitrogen-based explosive called dynamite. Then, in 1887, Nobel went on to invent a smokeless gunpowder called ballistite.

French chemist Éleuthère Irénée du Pont de Nemours (1771–1834), the founder of the giant chemical corporation Du Pont, revolutionized the manufacturing of gunpowder in the nineteenth century. One of Du Pont's major innovations was to make gunpowder using sodium nitrate. His process was much cheaper than producing gunpowders based on potassium nitrate.

A spectacular fireworks explosion lights up the sky. Potassium compounds were first used in fireworks displays in China over 1,000 years ago.

Periodic table

Everything in the universe consists of combinations of substances called elements. They are made of tiny atoms, which are too small to see. Atoms are the building blocks of matter.

The character of an atom depends on how many even tinier particles (called protons) there are in its center, or nucleus. An element's atomic number is the same as the number of its protons.

Scientists have found around 110 different elements. About 90 elements occur naturally on Earth. The rest have been made in experiments.

All these elements are set out on a chart called the periodic table. This lists all the elements in order according to their atomic number.

The elements at the left of the table are metals. Those at the right are nonmetals. Between the metals and the nonmetals are the metalloids, which sometimes act like metals and sometimes like nonmetals.

- On the left of the table are the alkali metals. All these elements have just one electron in their outer shell.

- On the right of the periodic table are the noble gases. All these elements have full outer shells.

- Elements in the same group have the same number of electrons in their outer shells.

- Elements get more reactive as you go down a group.

- The number of electrons orbiting the nucleus increases down each group.

- The transition metals are in the middle of the table, between Groups II and III.

Transition metals

Group I

Group II

Group I	Group II							
1 H Hydrogen 1								
3 Li Lithium 7	4 Be Beryllium 9							
11 Na Sodium 23	12 Mg Magnesium 24							
19 K Potassium 39	20 Ca Calcium 40	21 Sc Scandium 45	22 Ti Titanium 48	23 V Vanadium 51	24 Cr Chromium 52	25 Mn Manganese 55	26 Fe Iron 56	27 Co Cobalt 59
37 Rb Rubidium 85	38 Sr Strontium 88	39 Y Yttrium 89	40 Zr Zirconium 91	41 Nb Niobium 93	42 Mo Molybdenum 96	43 Tc Technetium (98)	44 Ru Ruthenium 101	45 Rh Rhodium 103
55 Cs Cesium 133	56 Ba Barium 137	71 Lu Lutetium 175	72 Hf Hafnium 179	73 Ta Tantalum 181	74 W Tungsten 184	75 Re Rhenium 186	76 Os Osmium 190	77 Ir Iridium 192
87 Fr Francium 223	88 Ra Radium 226	103 Lr Lawrencium (260)	104 Unq Unnilquadium (261)	105 Unp Unnilpentium (262)	106 Unh Unnilhexium (263)	107 Uns Unnilseptium (?)	108 Uno Unniloctium (?)	109 Une Unillenium (?)

Lanthanide elements

Actinide elements

57 La Lanthanum 139	58 Ce Cerium 140	59 Pr Praseodymium 141	60 Nd Neodymium 144	61 Pm Promethium (145)
89 Ac Actinium 227	90 Th Thorium 232	91 Pa Protactinium 231	92 U Uranium 238	93 Np Neptunium (237)

The horizontal rows are called periods. As you go across a period, the atomic number increases by one from each element to the next. The vertical columns are called groups. Elements get heavier as you go down a group. All the elements in a group have the same number of electrons in their outer shells. This means they react in similar ways.

The transition metals fall between Groups II and III. Their electron shells fill up in an unusual way. The lanthanide elements and the actinide elements are set apart from the main table to make it easier to read. All the lanthanide elements and the actinide elements are quite rare.

Potassium in the table

Potassium is an alkali metal, along with the elements lithium, sodium, rubidium, cesium, and francium. The alkali metals belong to Group I, which lies at the far left of the periodic table. These metals take their name from their reaction with water, forming alkalis (substances that neutralize acids) and hydrogen gas.

- ☐ Metals
- ☐ Metalloids (semimetals)
- ☐ Nonmetals

| | | | 20 K Potassium 39 | Atomic (proton) number / Symbol / Name / Atomic mass |

Group VIII

Group III	Group IV	Group V	Group VI	Group VII	Group VIII
					2 He Helium 4
5 B Boron 11	6 C Carbon 12	7 N Nitrogen 14	8 O Oxygen 16	9 F Fluorine 19	10 Ne Neon 20
13 Al Aluminum 27	14 Si Silicon 28	15 P Phosphorus 31	16 S Sulfur 32	17 Cl Chlorine 35	18 Ar Argon 40

			Group III	Group IV	Group V	Group VI	Group VII	Group VIII
28 Ni Nickel 59	29 Cu Copper 64	30 Zn Zinc 65	31 Ga Gallium 70	32 Ge Germanium 73	33 As Arsenic 75	34 Se Selenium 79	35 Br Bromine 80	36 Kr Krypton 84
46 Pd Palladium 106	47 Ag Silver 108	48 Cd Cadmium 112	49 In Indium 115	50 Sn Tin 119	51 Sb Antimony 122	52 Te Tellurium 128	53 I Iodine 127	54 Xe Xenon 131
78 Pt Platinum 195	79 Au Gold 197	80 Hg Mercury 201	81 Tl Thallium 204	82 Pb Lead 207	83 Bi Bismuth 209	84 Po Polonium (209)	85 At Astatine (210)	86 Rn Radon (222)

62 Sm Samarium 150	63 Eu Europium 152	64 Gd Gadolinium 157	65 Tb Terbium 159	66 Dy Dysprosium 163	67 Ho Holmium 165	68 Er Erbium 167	69 Tm Thulium 169	70 Yb Ytterbium 173
94 Pu Plutonium (244)	95 Am Americium (243)	96 Cm Curium (247)	97 Bk Berkelium (247)	98 Cf Californium (251)	99 Es Einsteinium (252)	100 Fm Fermium (257)	101 Md Mendelevium (258)	102 No Nobelium (259)

Chemical reactions

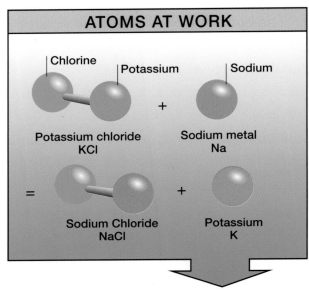

ATOMS AT WORK

Chlorine | Potassium | Sodium

Potassium chloride
KCl

Sodium metal
Na

Sodium Chloride
NaCl

Potassium
K

The reaction that takes place when potassium chloride reacts with sodium is written like this:

KCl + Na → NaCl + K

$\mathbf{C}$hemical reactions occur all the time. Some reactions involve just two substances; others many more. But whenever a reaction takes place, at least one substance is changed.

In a chemical reaction, the atoms stay the same. But they join up in different combinations to form new molecules.

A bright yellow precipitate (a solid) is formed during the reaction of lead nitrate (PbNO₃) and potassium iodide (KI) solution in water.

Writing down equations

Chemical reactions can be described by writing down the atoms and molecules before the reaction and the atoms and molecules after the reaction. The number of atoms before will be the same as the number of atoms after. Chemists write the reaction as an equation. This shows what happens in the chemical reaction.

Making it balance

When the numbers of each atom on both sides of the equation are equal, the equation is balanced. If the numbers are not equal, something is wrong. A chemist adjusts the number of atoms involved until the equation does balance.

Glossary

alloy: A mixture of a metal with one or more other elements.

atom: The smallest part of an element that has all the properties of that element. Each atom is less than a millionth of an inch in diameter.

atomic mass: The number of protons and neutrons in an atom.

atomic number: The number of protons in an atom.

bond: The attraction between two atoms, or ions, that holds them together.

catalyst: Something that makes a chemical reaction occur more quickly.

compound: A substance made of two or more elements bonded together. Potassium chloride is a compound that consists of one potassium and one chlorine atom.

crystal: A solid substance in which the atoms are arranged in a regular, three-dimensional pattern.

electrolysis: The use of electricity to change a substance chemically.

electron: A tiny particle with a negative charge. Electrons are found inside atoms, where they orbit the nucleus in layers called electron shells.

element: A substance that is made from only one type of atom.

fertilizer: A substances added to the soil to help plants grow better.

ion: An atom that has lost or gained electrons. Ions have either a positive or negative electrical charge.

metal: An element on the left-hand side of the periodic table.

molecule: A particle that contains atoms held together by chemical bonds.

neutron: A tiny particle with no electrical charge. Neutrons are found in the nucleus of every atom except hydrogen.

nonmetal: An element on the right-hand side of the periodic table.

nucleus: The dense structure at the center of an atom.

oxidation: A reaction in which atoms lose electrons.

periodic table: A chart of all the chemical elements laid out in order of their atomic number.

products: The substances formed in a chemical reaction.

proton: A tiny particle with a positive charge. Protons are found inside the nucleus of an atom.

reactants: The substances that react together in a chemical reaction.

redox reaction: A reaction in which both oxidation and reduction take place.

reduction: A reaction in which atoms gain electrons.

surface tension: An attractive force that pulls molecules in a liquid together.

surfactant: A substance that reduces the surface tension of water.

Index

For Carl ~ C F

For Sarah, Chris and Hannah ~ S M

This edition produced 2007 for
BOOKS ARE FUN LTD
1680 Hwy 1 North, Fairfield, Iowa, IA 52556

Copyright © 2007 by Good Books, Intercourse, PA 17534

Text copyright © Claire Freedman 2007
Illustrations copyright © Simon Mendez 2007
Original edition published in English by Little Tiger Press,
an imprint of Magi Publications, London, England, 2007

Library of Congress Cataloging-in-Publication Data
was submitted but was unavailable at the time of publication.

I Love You, Sleepyhead

Claire Freedman   Simon Mendez

Kate -
Merry Christmas
2007 :
We love you -
Uncle Roger
Aunt Annie
Cruz
+
Grace

Good Books

Intercourse, PA 17534
800/762-7171
www.GoodBooks.com

Look, little child,
as the night is unfurled,
The animals are going to bed
all around the world.

Close to her mother
and safe by her side,
Sweet little fawn
is so sleepy-eyed.

Nestled in grass,
as the soft breezes blow,
Bathed by the warmth
of the sun's evening glow.

Lion cubs romp as the sun slips away.
In the soft golden light, there's still time to play.

Soon they'll be yawning, three tired sleepyheads,
Watched by their mother all night in their beds.

Waddling to Mommy,
the tired ducklings quack,
Sleepy from swimming,
they're glad to be back.

Safely they're tucked
in their nest for the night,
Feathery bundles,
huddled up tight.

Daylight is fading fast, softly dusk falls.
"Bedtime, my little ones," mother fox calls.

"Mom, we're not sleepy!" the small foxes cry,
As low in the sky, the sun says goodbye.

Wrapped up in love,
little bear feels so snug,
Cuddled goodnight
in a big mommy-hug.

Drifting to sleep
he sinks into her fur,
Warm in the soft snow,
snuggled with her.

High up, the trees catch the last rays of sun,
As three tired monkeys climb up to their mom.

The sounds of the jungle, the rustling of leaves,
Lull them to sleep in the cool evening breeze.

Snug with their mommy,
the rabbits are all
Tumbled together
in one furry ball.

Cozy and warm,
they will sleep safe and sound,
Curled in their bed
on the soft mossy ground.

Rocked by the waves
beneath velvet blue skies,
Wrapped in her mommy's arms,
small otter lies.

Under the stars
in the dappled moonlight,
"One kiss," smiles Mommy,
"and then it's sleep-tight."

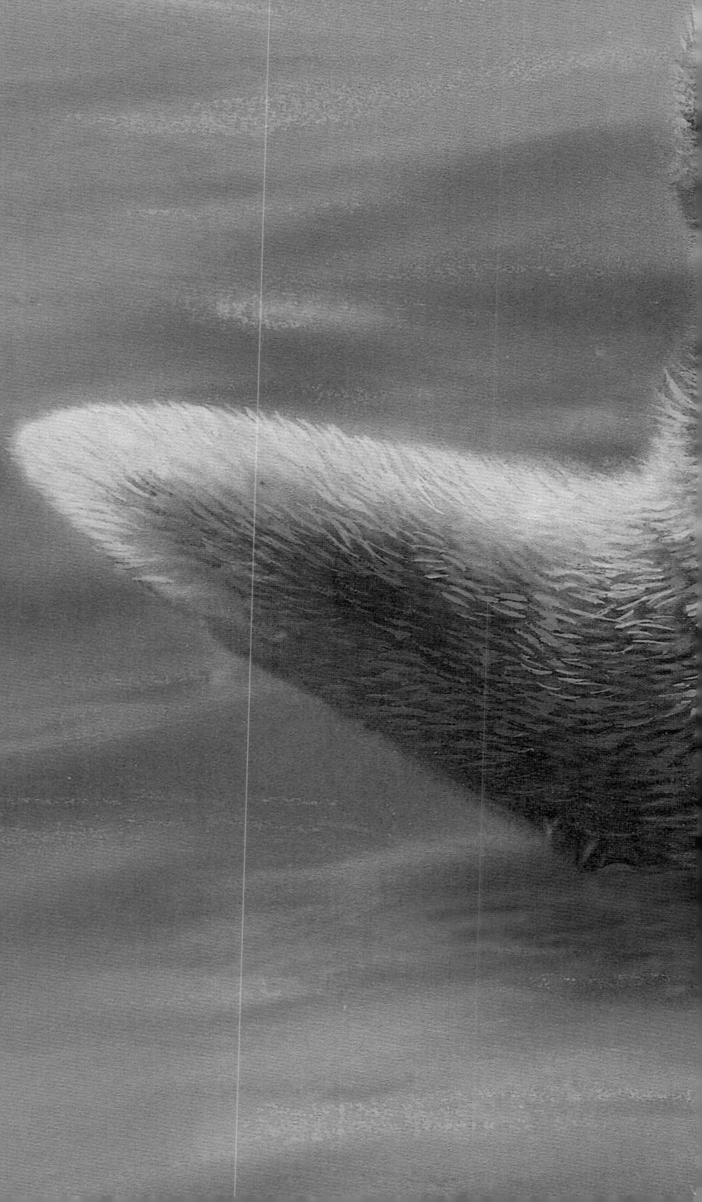

Snowflakes are swirling,
all fluffy and white,
Sparkling like stars in the
gleaming moonlight.

Cuddled up close,
little penguin stays warm,
Through the cold frosty night,
till the first light of dawn.

As mother owl hoots
her sweet, low lullaby,
Her baby owls blink
at the star-studded sky.

Through the dark treetops,
her echoing call,
Sings to the world,
"Goodnight to you all!"

Baby whale drifts
to the deep ocean's song,
Close to his mother,
all the night long.

Down through the water,
the soft moonlight streams,
As little whale floats
in a sea of sweet dreams.

Small panda sleeps
as the stars peek-a-boo,
Held by his mother,
all the night through.

Cuddled up close,
she gives him a kiss.
Tucked in together,
they're perfect like this.

Sleep, my child, sleep,
'neath the moon's silver light.
I love you, sleepyhead,
sweet dreams—goodnight!